Renard Teipelke

Facebook and Travel Mobility

The Interrelation between Social Network Services and Physical Travel and Its Possible Implications for the Future

GRIN Verlag

Bibliografische Information der Deutschen Nationalbibliothek:

Die Deutsche Bibliothek verzeichnet diese Publikation in der Deutschen National-
bibliografie; detaillierte bibliografische Daten sind im Internet über http://dnb.d-
nb.de/ abrufbar.

Imprint:

Copyright © 2012 GRIN Verlag GmbH
Druck und Bindung: Books on Demand GmbH, Norderstedt Germany
ISBN: 978-3-656-53893-6

This book at GRIN:

http://www.grin.com/en/e-book/264477/facebook-and-travel-mobility

GRIN - Your knowledge has value

Der GRIN Verlag publiziert seit 1998 wissenschaftliche Arbeiten von Studenten, Hochschullehrern und anderen Akademikern als eBook und gedrucktes Buch. Die Verlagswebsite www.grin.com ist die ideale Plattform zur Veröffentlichung von Hausarbeiten, Abschlussarbeiten, wissenschaftlichen Aufsätzen, Dissertationen und Fachbüchern.

Visit us on the internet:

http://www.grin.com/

http://www.facebook.com/grincom

http://www.twitter.com/grin_com

Veranstaltung:	Modul HG2 Kultur und Gesellschaft: Seminar „Mobilität 2.0 – Soziale Veränderungen und Zukunft der Mobilität" (WiSe 2011/2012)
Abgabetermin:	26.03.2012
Autor:	Renard Teipelke
Studiengang:	M.A. Geografien der Globalisierung

Facebook and Travel Mobility: The Interrelation between Social Network Services and Physical Travel and Its Possible Implications for the Future

Table of Content

Abstract

In an era of information and communication technologies, imaginative and virtual travel have not substituted physical travel but resulted in a network society of multiple mobilities. People are increasingly mobile for maintaining their professional and private social relationships in intermittent face-to-face meetings. So far, we do not know much about the interrelation between travel mobility and social network services (SNSs). With the example of Facebook as the dominating, undisputed SNS of 845 million monthly active users, I will argue how 'place'-focused features are put under the spotlight of Facebook users' activities and perceived character. As a platform combining various services, tools, and applications also of third parties, Facebook can influence its users' mobility habits and expectations of physical traveling. This brings with it important implications for research and practice, since foremostly young generations are concerned and will thus influence future developments.

1. Introductory Remarks

A friend of mine has just updated her status on Facebook, posting that she arrived in Kuala Lumpur for her spring holidays. My colleague changed his Facebook profile, posting that he is going to work with a micro finance bank in Nairobi next month. And then, I am stumbling across my cousin's photo album of his last family reunion somewhere near Lake Tahoe.

With its 845 million monthly active users at the end of 2011 (Facebook 2012: 1), Facebook (www.facebook.com) is now the dominating, undisputed social networking website in the world. And from typical comments such as the ones above, it seems obvious that place or location plays a highly relevant role on Facebook. Where you come from, where you currently are, and where you will be is a major feature of characterizing a Facebook user.

This is related to mobility, because being in a place often implies going there first. Thus, people are traveling. And they are traveling at a constantly increasing rate. Szerszynski and Urry (2006: 116) usually compare the 25 million legal international tourist arrivals in 1950 with the 760 million of 2004 and the expected 1.6 billion in 2020. It goes without saying that the travel and tourist industry, therefore, has a fundamental stake in the world's gross domestic product, exports, and employment (ibid: 116). The reasons for this increased travel mobility are manifold – covering areas of leisure and business traveling to formal migration and illegal trafficking. At this juncture, we have to differentiate travel mobility from other forms of mobility – a widely used term describing various aspects. For this paper, the focus lies on personal travel mobility as the physical 'movement' of people from one place to another of a particular distance and for reasons beyond simple daily practices such as going to work, the supermarket, or the cinema in one's hometown. Thus, this mobility always includes a specific social element in its intention to travel to a certain place at a certain time to maintain social relationships or to experience this place on one's own (cf. Sheller and Urry 2006: 217).

It might seem obvious that the interrelation of increased travel mobility and social network services (SNSs) should have been a focus in previous research. But quite the contrary, the current state of the art has neither looked closer into the effects of SNSs on travel mobility, nor the effects of travel mobility on these SNSs (cf. Ohnmacht 2009: 120-121). We have a good understanding of what online activities SNSs are triggering, but we have hardly studied the influence of SNSs' uses on offline activities, travel mobility in particular (cf. Barkhuus and Tashiro 2010: 133).

Therefore, I will take up this promising research perspective by introducing Facebook's 'place'-focus and discuss the interrelation of travel mobility and this SNS, particularly its young user base and possible implications for the future. My main argument is that Facebook offers features (services, tools, applications etc.) that are putting 'place' under the spotlight of a user's activities, thus constituting at least indirectly the perceived character of a user. So to say, Facebook is much about where you have been/are/will be. With the constant communication of various places on a user's Facebook page, virtual and imaginative travel are not any longer satisfying to his/her friends, but are forming an understanding of, piquing a curiosity for, and triggering physical travel beyond previous ideas – places that have been (too) far away in the past are now increasingly becoming travel destinations one has to visit sometime in the (near) future. As a platform combining various other services and applications in the field of travel mobility, Facebook will influence its users' mobility habits. This development, especially since it concerns foremostly younger generations, can have tremendous impacts on transportation infrastructure systems. This calls for an approach in research and practice that eventually realizes the "mobility turn" (Sheller and Urry 2006: 208), which will be further explained in the paper.

Since Facebook is the combined result of what its producers are programming and what its users are making out of these services, tools, and integrated applications of third parties, it is particularly challenging to describe a current development which outcomes are hard to predict in this fluid setting. Therefore, we can only look at relevant aspects of Facebook and try to understand possible interrelations step by step. In order to grasp the topic logically, I will briefly cover previous research on mobility and SNSs (chapter 2). Then, I will explain why I chose Facebook as a research object (chapter 3.1.) and introduce its relevant ('place') features (chapter 3.2.). These chapters will function as the foundation for discussing the interrelation between Facebook and mobility with regard to modern-day traveling (chapter 4.1.), explanations for traveling (chapter 4.2.), Facebook's role in travel mobility (chapter 4.3.), and the related impact on various users/people (chapter 4.4.). Afterwards, I will indicate important implications for research and practice (chapter 5), before bringing the main findings together in a brief summary (chapter 6).

2. Mobility and Social Network Services in Research

Relevant contribution of research on the interrelation between mobility and SNSs will be taken on later in this paper. For the moment now, it shall be sufficient to briefly frame the topic with regard to the most important findings in previous studies. An important basis for comprehending the interrelation between SNSs and mobility is the differentiation into physical, imaginative, and virtual travel – referring to traveling with the body (such as an actual holiday trip), through images/media (such as a TV documentation), or via communication technologies and corresponding information (such as Google Earth), respectively (Szerszynski and Urry 2006: 115-116). While previous studies thought of non-physical traveling as a form that would increasingly substitute physical traveling, more and more researchers could show that we are actually dealing with multiple mobilities, which are interdependent and accelerate people's mobility (Larsen, Urry, and Axhausen 2006: 54-55; Urry 2003: 117; Sheller and Urry 2006: 212).

This notion related to research approaches has been framed as the "mobility turn" or "mobilities paradigm" by authors such as Urry (2007: 18-20, 44-60). The idea is that places are connected with each other in networks that are stretching far beyond a particular place. In order to understand these network connections, Sheller and Urry call for a new research approach in which we are "putting social relations into travel and connecting different forms of transport with complex patterns of social experience conducted through communications at-a-distance" (2006: 208). The modern world and thus everyday life are now organized, managed, and coordinated through new technologies that are, thus, also influencing (transport) mobility and social relations (Sheller and Urry 2006: 207).

In the same vein, Sheller and Urry (2006: 216) or Axhausen (2006: 3), amongst others, prominently explain the important role of weak ties in a network society. As Axhausen (ibid.) focuses on travel activities as a form of building and sustaining social capital, Urry (2007: 36-37) also argues how face-to-face relationships with people, places, and events necessitate traveling.

Various studies show the role of SNSs, such as travel websites (Germann Molz 2006), in these new interrelations between network connections and travel mobility – mediated or enabled through information and communication technologies. It is important to note that the traditional differentiation of various traveler types is no longer feasible as Larsen, Urry, and Axhausen explain in the following quote that can also function as a first summary before I introduce Facebook and its ('place') features:

> *"Tourist-type travel enters the lives of business people and global professionals, second-homeowners and their friends and families, exchange students and gap-year workers abroad, migrants and people with friends and families in distant places. Tourism is less the privilege of the rich few, but more something involving and affecting many people, as otherwise immobile people might occasionally visit or host distant kin or be heartbroken when they remain at-a-distance." (2006: 43)*

3. Facebook and Place

3.1. Facebook as a Research Object

A few years ago, authors always had to explain why they picked Facebook as a research object when talking about SNSs, since there have been many SNSs of similar size and reach. I will also briefly explain my choice in favor of Facebook; however, it will be clear that this explanation will probably become increasingly obvious in the future.

With 845 million monthly active users (Facebook 2012: 1), more than 30 billion pieces of shared content per day (Wauters 2011), and 770 billion page views each month (Burbary 2011), Facebook is one of the heavy weights not only in the SNSs, but in the internet in general. Other SNSs are still playing a somewhat relevant role only in a few high-populated countries, such as Russia (*V Kontakte*), China (*QZone*), and partly Brazil (*Orkut*) (Wauters 2011). Facebook is the number one SNS in 119 out of 134 countries of the world in 2011 (ibid.), and in many countries a large amount of users is registered on Facebook (United States more than 155 million, India, Indonesia, and Brazil with more than 42 million, European countries like Great Britain, France, Germany, and Italy with more than 20 million users) (Socialbrakers 2012c). Furthermore, countries like the United Arab Emirates, Norway, Chile, or Taiwan have a penetration rate of Facebook users of more than 50 percent (ibid.). And in the past six months, countries such as Brazil, India, Mexico, Japan, Indonesia, or Germany contributed several million new users to Facebook (ibid.).

I am outlining these statistics not only to show why Facebook is a relevant research object, but also to argue that the interrelation between SNSs and (offline) mobility can only or best be understood if we study Facebook as the undisputed and still constantly increasing SNS worldwide – both in terms of scale (as the statistics show) and scope (as the following introduction of Facebook's (new) features will underscore).

3.2.   The Role of Place on Facebook

Since various updates to its services, tools, and applications have been undertaken in 2011 and early 2012, I cannot fully explain the social network service(s) of/on Facebook (www.facebook.com). Therefore, I will highlight 'place' features. This is already apparent in the rather simple *profile* information, users can and often do provide with regard to their current city, workplace, their high school and university (or other educational institution), and their hometown. The *open-graph* of the newly introduced *timeline* as a form of "remote-control autobiography" (Levy 2011) summarizes and archives a user's 'biography' (*life events*) and activities on Facebook and affiliated/interconnected services chronologically and in order of relevance (depending on fine-tuned algorithms as well as a user's individual profile settings). Interestingly, these pieces of information often include very precise dates and locations, which might be explained by the importance time and place play in many activities (ranging from when and where I was born to when and where I graduated etc.).

A user can search her/his list of Facebook *friends* by current city, school, workplace, hometown, and interest. These categories hint to the role location plays on Facebook.

5

Friends and places become especially relevant with regard to two Facebook features: the *status update* and *photos/videos*. Updating your *status*, you can write anything you want, but there are specific options that facilitate an easy *check-in* of friends who are currently with you and of the place you are currently in. While the previous tool *places* was a separate function, the *location check-in* feature is now integrated in the *status update* – and in other information fields such as the caption of a photo in your album (Fiveash 2011).

This leads us to *photos/videos* on Facebook. More than 250 million photos are uploaded on Facebook per day, making the sharing of photos one of the most popular activities of Facebook users (Shaffer 2011). The tool has been improved by various product updates since 2010 (ibid.), and the new profile (*timeline*) has now a big *cover picture* at its top. Anyone who uses Facebook might have already recognized that uploaded photos as well as the cover picture of profiles in many cases show people (friends) and places which then also often get tagged by the user (Facebook, for instance, asks a user when uploading a photo: Who were you with? Where was this picture taken? – and offers direct tools to tag a friend or to check-in a date and place for this photo).

The 'place' focus is further emphasized by a *map* tool integrated in a user's profile showing the world map with marks of places where the user has been some time ("Places Lived", "Trips", "Life Events") or did a lot of Facebook-related activities (such as uploading photos of the last trip to Buenos Aires).

I will get back to other Facebook features later, so I only briefly mention now that the *news feed* is something like the *home* page of Facebook where a user can see all or only a selected kind of *updates* of friends and *fan pages* s/he liked (the *Like*-button is another well-known Facebook feature). The *ticker* is a split-screen on the right side of the Facebook page through which constant short-cut information of friends' activities are rolling. Facebook users can create *events* or *groups* for various purposes (for instance, planning the next kayak trip to Scandinavia or discussing questions around a graduate class field trip to Pakistan).

Where Facebook really came up with innovations (especially in the field of micro-targeted ads) has been the integration of ample *applications* for other services, ranging from news (Washington Post Social Reader) and media (Spotify) to games (Farmville) and advice platforms (TripAdvisor). With more than 20 million app-installations per day (Pring 2012) and improving mobile devices (O'Farrell 2012), Facebook is used more often and increasingly becomes a platform that combines, coordinates, and connects many other SNSs and applications as well as archiving and analyzing users' activities (and interests etc.).

While it is quite difficult to describe the use and the experience of using Facebook in words, I can recommend watching three advertising videos of Facebook introducing its main new features, such as the *timeline*, *friends lists*, *news feed*, *ticker*, and *apps* (Facebook 2011a, b, c). Although the three videos in sum run less than four minutes, the following places are shown in various Facebook tools: Palo Alto, Japan, a beach, the Valley (CA), Alaska, Niagara Falls, Marilla (NY), Buffalo, San Francisco, and the Golden Gate Park. We can also recognize that other aspects are relevant on Facebook as well: family and friends, education, music, cooking, and sports. Nevertheless, place is one of the major defining features.

4. Synthesis of Travel Mobility and Facebook

4.1. Traveling in the 21st Century

"Transport and communication technologies are thus 'travel partners', components of 'network capital'. We might see this as a process of co-evolution, between new forms of social networking on the one hand, and extensive forms of physical travel, now normally enhanced by new communications, on the other. These sets of processes reinforce and extend each other in ways that are difficult to reverse." (Urry 2007: 179)

This quote perfectly grasps the interrelation between travel mobility and SNSs in our modern times. In addition, we have to explain how and why this phenomenon happens. I will deal with explanations for traveling in the next chapter (chapter 4.2.), but first address the 'how' question.

Ellison, Lampe, and Steinfield (2009: 8) describe how SNS users create "hyper-local, ad hoc networks" that are blurring the line between offline and online interaction. This is also the case with regard to physical travel and SNSs. It already starts with the organization of trips when SNSs are used as a platform to seek advice, receive feedbacks or comments, and use applications on Facebook such as TripAdvisor to plan accordingly. Rhodes (2010) describes how the – we might want to use the term – 'traveler 2.0' keeps friends on Facebook constantly informed about her/his location and status. This real-time reporting also functions as a tool to share experiences, information, and images. Without mobile devices and corresponding information technologies this kind of "interactive travel" (Germann Molz 2006: 378) would not be possible.

Germann Molz (2006: 378-380) explains the other aspect to this type of traveling: the audience. The author sees SNSs offering surveillance tools through which users/travelers can perfectly engage their audience. It is particularly the feedback loop (ibid: 389) that keeps the interconnection between traveler and audience, offline and online activities, alive, since the SNS audience is asking for updates, images, reports etc. and the traveler is willing to provide these information pieces.

While Germann Molz's work (ibid.) is helpful to comprehend the role that travel reporting and the corresponding interrelation between travel mobility and SNSs plays, the recent developments on Facebook hint to further aspects in this interrelation: Surveillance or imaginative traveling with the information-providing user/traveler seems to be no longer sufficiently satisfying. The audience also wants to experience what the traveler reports by undertaking their own physical travel. Before I further elaborate on this argument (chapter 4.3.), we have to understand two possible explanations for why people are actually traveling.

4.2.   Explanations for Traveling

One simple and straight-forward explanation for traveling is that people want and even need to meet other people from time to time in order to maintain their various networks (Urry 2003: 161; Urry 2007: 36-37). This underscores why the differentiation between leisure and business travel, for instance, is not very helpful any longer (Sheller and Urry 2006: 212). It could be your friend or a family member, but also a colleague or client – traveling and intermittent face-to-face meetings are necessary for our professional as well as private network connections (Axhausen 2006: 5). Some authors (Sheller and Urry 2006: 217) even go further when noting that these direct conversations and meetings are not only necessary but also bound to particular places and moments – for instance a family reunion does have to happen on Christmas in the grandparents' old house in the original hometown. Even though the meeting of friends and kin is a personal social reason for traveling to a certain place at a certain time, the examples given above result in a rather functional explanation for traveling. It is not the traveling that is important but the meeting in the place of destination.

Therefore, we have to add another explanation for traveling. We can best grasp this experience-focused explanation by turning to Szerszynski and Urry's notion of a "culture of cosmopolitanism" (2006: 114). They explain this idea with regard to the increased and multiple mobilities of people. This cosmopolitanism (ibid: 114) includes an "extensive mobility" in terms of physical, imaginative, and virtual travel. It can be described as a "curiosity about many places, peoples and cultures", the "willingness to take risks by virtue of encountering the 'other'", and the "ability to 'map' one's own society and its culture in terms of a historical and geographical knowledge" (ibid: 115). Other aspects can include the "semiotic skill to be able to interpret images of various others" and the "openness to other peoples and cultures" (ibid: 115). People traveling as 'cosmopolitans' might include a business meeting in their trip, but they are traveling foremostly to experience a particular place for themselves (Urry 2003: 162-164). It is even very common that the traveling as the time spent moving has a particular meaning (catchphrase: traveling without destination). We can relate this idea to Germann Molz's explanation of traveling as a personal transformation

8

or growth, as a process of sensing and learning (2006: 388-390). Thereby, curiosity for the new or different plays an important role.

4.3. How Facebook Makes a Difference

Both explanations for traveling play a relevant role with regard to SNSs such as Facebook. I have already explained how traveling in the 21st century combines online and offline activities through information and communication technologies as well as network connections (chapter 4.1.). I have also introduced various features of Facebook that are putting 'place' under the spotlight of users' activities and perceived character (chapter 3.2.). If we bring these aspects together, we can understand my main argument that surveillance by an audience is no longer sufficiently satisfying and that the audience/users want to experience places they got to know on Facebook through their friend's travel reporting. Or to put it another way: In order to report on Facebook about special places and trips a user experienced, the user needs to physically travel in the first place. Urry (2002: 269) refers to this explanation for traveling also indirectly in his studies when discussing how a person might eventually want to physically travel to a certain place, the more s/he has seen/heard/consumed this place through imaginative or virtual travel. Urry furthermore listed "event obligations" (2003: 164) as one reason for traveling, which he describes as events that necessarily need to be experienced "live". I will now explain how and why Facebook becomes increasingly important with regard to the interrelation between SNSs and mobility – and increased traveling as I am arguing.

The newly introduced features on Facebook (such as the *timeline* with its *open-graph*, integrated *applications*, or the *check-in* of friends and places in a user's *status update*) can all be summarized under the motto 'frictionless sharing and experience' (cf. Ingram 2011). Claims about Facebook's hidden agenda to make money with users' personalized data and concerns about an "inevitable privacy backlash" (ibid.) are all relevant and worth a discussion, but are not the focus of my paper. However, related to these critical aspects is Facebook founder Mark Zuckerberg's 'law of social sharing' (ibid.), in which he once stated that users will double the amount of data they are sharing with SNSs such as Facebook every year. Until now, he has been pretty correct with this prediction (Pring 2012).

This extensive data sharing is met with well-designed algorithms and corresponding tools and applications on Facebook (especially the *timeline* and the *ticker*) that 'help' users not only to experience their friends' information and activities in a frictionless style, but also to 'accidentally' find and recognize similarities in their interests. Facebook has termed these two key elements of the new features "real-time serendipity" and "finding patterns" (Ingram 2011). Serendipity can be described as the discovery of something relevant to you while seeking something different or just distraction. Therefore, Facebook (and other SNSs as well) proves that social networks are about discovering your friends and (their) interests, recognizing similarities or patterns, and, thus, maintaining and reinforcing the network connections between you and your friends, as well as concrete information and leisure time interests (cf. Levy 2011).

This means that even lightweight pieces of data, such as photos of a user's last trip or the check-in of places in her/his status update, can be very meaningful to users and their friends. Applied to travel mobility, Axhausen (2006: 5) assumes that people who are more active in SNSs as well as traveling have a broader "personal world" and develop greater expectations of other places they still want to physically experience. As my argument goes, I would even assume that a highly mobile Facebook user's (frequent traveler's) expectations can trigger her/his friends also to develop stronger curiosity and greater expectations concerning future travel destinations as Facebook puts 'places' under the spotlight: It is about where you have been/are/will be. And *frictionless data sharing and experience*, *real-time serendipity*, and *finding patterns* show Facebook users how places that might have been perceived in the past as (too) far away from one's own place are now reachable – as the 'travel reporting' of your Facebook friends proves.

Referring back to the explanations for traveling, Facebook even seems to combine them by giving its users the opportunity to *check-in* friends and places that can be colleagues as well as kin, a business meeting place as well as a beach, thus ascribing traveling both the functional and the experience-focused explanation. The underlying idea of multi-purpose trips is not new (Axhausen 2006: 7). However, where Facebook makes a difference is the enhancement of its users' imagination or expectation that many places are now possible and worth to be discovered.

We can find clues on Facebook that travel-related aspects are playing a fundamental role in this social network: More than 12 million users have listed traveling as their *interest* – compared to the more than 12 million who like cooking, the more than 16 million who like sleeping, or the more than 15 million who like football (soccer), traveling is part of the top ranking interests (www.facebook.com). *Fan pages* of travel destinations have a large fan crowd, such as the entertainment park Disneyland with more than 13 million fans, countries like Australia or Colombia with around 2 million fans, cities like Paris or Berlin with more than 1 million fans, or museums, clubs, and particular hotels with several hundred thousand fans (Socialbrakers 2012a, b). I have already pointed out that more than 250 million photos are uploaded each day on Facebook (Shaffer 2011) – through the new features often with tagged places. It might be an explanation mark behind the place and travel focus of Facebook and its users' activities that TripAdvisor as the market leader in travel advising has around 16 million active users per month on its Facebook-integrated *application*

(Socialbrakers 2012d). Taking into account that these applications, services, and tools have been introduced only recently, their 'explosive' increase in active users hints to possible developments in the near future.

Facebook makes a difference in comparison to other offline and online services in that it functions as a platform that combines *user-generated content* (updates, photos, recommendations, and comments etc.) with *integrated* or *interconnected applications* of other service providers (such as TripAdvisor) and *micro-targeted ads* (featuring, for instance, cheap online tickets of a low-cost airline for a Facebook user who's workplace is close to a peripheral airport), thereby conflating frictionless sharing and experience of friends traveling with the organization of the next trip as a 'normal' interrelated process of using SNSs.

4.4. For Whom Facebook Makes a Difference

Facebook's role in the interrelation between mobility and SNSs is also a specific one because of Facebook's young user base (Pring 2012). This can be related to previous research in which authors have often highlighted different mobility habits of younger people. Ohnmacht (2009: 111, 193) shows how people between 15 and 35 are highly mobile and have more distant social connections (particularly outside of their hometown). Larsen, Urry, and Axhausen (2006: 128) also underscore the high mobility of young adults, in their case study specifically people before the child-rearing part of their lives. Mau and Mewes (2009: 180) discuss transnational experiences of having become very much part of our everyday life and how the virtual or physical crossing of borders has become natural in the course of maintaining one's own network connections – a notion on which Urry (2007: 208-209) also indirectly hints when pointing out the increased traveling of young adults.

These different mobility habits of younger people can be explained by mobility biographies to a certain degree. Lanzendorf (2003: 8), for instance, argues that the travel behavior is too a large extent habitual. Life course events can necessitate place changes, due to family, education, job etc., and can influence mobility habits (ibid: 11). Since young adults are particularly disposed to such life changing events when moving to a different place for their university education, or when being (forced to be) more mobile for their tasks in a new job, or when visiting their parents during the holidays in their hometown, their mobility habits are exposed to many possible changes (cf. Ohnmacht 2009: 53).

In such a situation, information and communication technologies, in particular Facebook, function as tools to plan, coordinate, and report on trips (for instance vacation), meetings (such as business fares), events (for example New Year's Eve) etc. Furthermore, SNSs help young adults to maintain their network connections (cf. Barkhuus and Tashiro 2010). Since studies have shown how young adults are heavy users of SNSs (Ellison, Lampe, and

Steinfield 2009: 6) and early adopters of new technologies (Barkhuus and Tashiro 2010: 134), current generations of people between 15 and 35 are growing up with Facebook as their major 'system' to organize their interrelation between mobility/physical travel and social relations. To exemplify the scope of this development, we only have to be aware that close to three quarters of all Facebook users are younger than 35 (Burbary 2011) and that there have been more than 100 billion friend connections on Facebook at the end of 2011 (Facebook 2012: 1), implying necessary intermittent face-to-face meetings (cf. Axhausen 2006: 5).

This generation of young Facebook users goes beyond conclusions in previous studies: There has often been made the claim that high mobility or even hyper-mobility foremostly concerns industrialized societies of the Western hemisphere (cf. Urry 2003: 159; Urry 2002: 258). Furthermore, results from qualitative and quantitative studies have shown how better educated and higher status people are more active in transnational networks, and thus more mobile (cf. Mau and Mewes 2009: 180; Ohnmacht 2009: 193). I will not refute these findings, but I think that this picture of mostly Western, well-off, highly educated young adults being increasingly mobile is still too limited.

I have already presented statistics that show how deeply Facebook reaches into developed as well as developing countries (chapter 3.1.). Besides, we have to go beyond the 'internationally connected highly mobile SNS-active student' and take into account other settings that put people (or young adults in the case of Facebook) in a situation that might trigger or even necessitate higher mobility (cf. Szerszynski and Urry 2006: 119). I can imagine that the focus on students and/or people in industrialized countries in previous studies has been caused by a limited perspective on Anglo-Saxon education and job systems and the corresponding places that were studied (particularly the United States and Great Britain). If we take two constructed examples, the case of a German auto mechanic apprentice and the case of a young Taiwanese cook working abroad, we can get a better understanding of this new generation of highly mobile people.

In the first case, we have an apprentice being trained as an auto mechanic in Germany. This person is not a student but very likely to earn enough money in his/her future job in order to afford world-wide traveling. With Germany's high penetration rate of Facebook users (approximately one third of the country's total population, probably much higher penetration for young adults) (Socialbrakers 2012c), we can assume that the apprentice is using Facebook and sharing information and interests with friends. Even though the apprentice might not be connected to an international friends' network as a student who has studied a year abroad might be, the apprentice would nevertheless be exposed to Facebook's 'place' features, including the *map*, *photos*, and *applications*, such as TripAdvisor. One could question if the socialization through a university education and the mobility requirements that come along with a university-degree job would have a stronger impact on a person's travel mobility habits than the socialization through SNSs. Referring back to the experience-focused explanation for traveling, we could not say why the apprentice would necessarily be less mobile than the 'stereotypical student' of previous studies.

In the second case, we have a young Taiwanese cook working abroad. From previous studies, we would conclude that his education and occupation do not make him a highly mobile person or a likely 'global traveler'. However, the young cook does probably get a comparatively high salary from which he transfers money back to his family or kin in Taiwan. He might afford for his child a high-class education in another region or country. Via modern information and communication technologies, he can sustain his network connections – But as Urry (2003: 161; 2007:36-37) explains with respect to face-to-face meetings, the young cook would fly home once a year for his family's traditional holiday. With Taiwan's Facebook penetration rate of more than 51 percent (Socialbrakers 2012c) and the cook living abroad, he could probably be seen as a young adult with high physical, imaginative, and virtual mobility. If his child is going to study somewhere else in the future and other kin are also working abroad, it is very likely that his family and friends use SNSs like Facebook and report on their various experiences in different places as well as visit each other there.

What I want to show with these two constructed (not so unrealistic) examples is that previous studies that analyzed most often students and people in industrialized countries of the Western hemisphere have drawn conclusions from a limited perspective that does not allow predictions on the interrelation of SNSs and travel mobility in the future. A rather general notion by Ohnmacht (2009: 202) seems to fit much better: He argues that the size of the network geography and corresponding mobility of a person increases with her/his mobility biographical events – this leaves enough room for (young) adults of various educational backgrounds, occupations, and regional origins, as the previously quoted note by Larsen, Urry, and Axhausen (chapter 2) already implies. Therefore, we have to recognize that we are dealing with a highly diverse group of (young) adults using Facebook (or other SNSs) and being influenced in probably changing their travel mobility habits – this has direct implications for future research approaches and transportation infrastructure systems.

5. Implications for Research and Practice

The previous chapters have touched on various aspects of the interrelation between travel mobility and SNSs that are still little researched. With regard to future studies, the perspective needs to include people of different backgrounds (origin, race, ethnicity, education, occupation, family status etc.) (on previous limitations in studies cf. Cheung, Chiu, and Lee 2011: 1341-1342). Because of its dominating, undisputed, and increasingly embracing role, Facebook has to be the major research object. What Barkhuus and Tashiro criticized already in 2010 (133-135) still holds true: Many studies showed what further online activities are triggered by/on a SNS; what we need to do is to look at offline activities triggered by SNSs and vice versa. In our particular case, we would have to track Facebook activities of sharing data (*status updates, photos, check-in* of friends and places, *application* uses) with offline activities in the field of travel mobility. I have already argued why Facebook could increase physical travel, but for future research, we would have to test why SNS users decide to travel somewhere and if their SNS activity leads to an increased number of trips in comparison to previous years (longitudinal study) and non-SNS-users (cross-sectional study) (on challenges regarding the research design cf. Axhausen 2006). In order to understand possible interrelations between Facebook's 'place' features and physical travel, it could be promising to analyze the tagging of places in a user's friends network and corresponding follow-up activities of people in this network that are related to previously tagged places and, thus, point out possible patterns.

Even though I can only provide a personal anecdote to this research idea, it exemplifies the importance of tracking patterns and interconnections: In 2009, a couple of my friends went on a Thailand trip and reported on it on Facebook (for "interactive travel" cf. chapter 4.1.). After they posted the photos on Facebook, various other friends in our network started to discuss this trip – the photos seemed to show us how amazing the place was and they took us on an imaginative travel to Thailand. A few months afterwards, another group of my friends went to Thailand, also reporting on their trip with photos that pretty much enforced the previously imagined picture we had in mind of Thailand. Until last year, two other groups of friends undertook a similar trip. In the end, various discussions and follow-ups about the Thailand trips accompanied Facebook activities – even though the travel groups did not all know each other in person. It has been the photos and surrounding online activities that put a trip to Thailand under the spotlight in my friends' network. Even more: While a Thailand trip, including the Ko Phi Phi Lee island from the Hollywood movie "The Beach" (director: Danny Boyle, 2000) was unimaginable for most of our friends a couple of years ago, it has changed into a trip that is feasible. Furthermore, one has to undertake it sometime in order experience the amazing beaches etc. on one's own. In this example, a travel pattern or similarity in my friends' network becomes apparent and seems to have triggered physical travel or at least the curiosity and expectation of the audience to experience Thailand in a future trip. This all happened before Facebook introduced its new ('place'-focused) features. Therefore, it underscores how interesting and telling the above-outlined research approach could be in 2012 and the following years.

Beyond research questions, Facebook in its role of probably amplifying travel mobility can also have tremendous impacts in practice. I will not deal with possible implications for various transportation and tourist industries and businesses, but focus on political implications with regard to transportation infrastructure systems. At this juncture, it is well known that transportation costs have decreased and a further globalizing world economy necessitates higher mobility (Szerszynski and Urry 2006: 116; 210-211). There are more people traveling – for reasons of survival, occupation, education, social relations, or (self) experience (ibid: 116). Political stakeholders will have to deal with the role of information and communication technologies and SNSs in particular in an age of a network society (cf. Ohnmacht 2009: ix). If Facebook functions as a platform for planning, coordinating, and reporting on physical travel, and at the same time combines various other services and applications in the field of mobility and travel, it will influence its users mobility habits and expectations of physical traveling.

It goes without saying that an ever increasing mobility and more extensive network connections that necessitate intermittent face-to-face meetings in distant places stand against any objectives of a (more) sustainable development of transportation infrastructure as well as related political agendas concerning global warming, environmental degradation, or regional inequalities (Khisty and Zeitler 2001; Ohnmacht 2009: 214; Axhausen 2006: 6-7). This is especially critical with regard to what Adey, Budd, and Hubbard termed "aeromobility" (2007: 774), referring to flying as the new normal mode of international traveling. No matter if we are dealing with a "hypermobility" (ibid: 774) or 'solely' high mobility, it is necessary both in research and practice to eventually realize the "mobility turn" (Sheller and Urry 2006: 208) by bringing together knowledge about transportation infrastructure and mobility with SNSs research and travel studies.

6. Concluding Remarks

We can tell from the past five years that SNSs are now an integral part of our daily lives, and more and more offline as well as online activities are dependent on SNSs (Ellison, Lampe, and Steinfield 2009: 6) – in particular Facebook because of its size and reach. Various studies have shown that information and communication technologies do not substitute physical travel but enhance the mobility of people (Larsen, Urry, and Axhausen 2006: 46, 54-55). More and more generations grow up with SNSs as their major tools to maintain network connections that are characterizing our society. Assuming that the innovation will further improve services, tools, and applications of Facebook etc., the impact of SNSs on mobility habits and offline activities in general might well be far beyond our imagination and possibly affect our daily lives rather indirectly and pervasively. This makes it highly difficult to grasp research objects like Facebook. They perform various roles, offer plenty of uses, and are fluid in that their users decide in which direction the SNSs and related activities develop.

Even at the moment now, Facebook is not yet as mobile as its developers and users would like it to be. If mobile devices and infrastructures improve, not only Facebook but also its users will become more mobile (Barkhuus and Tashiro 2010: 135-136). The 'place' focus is not the only major element of Facebook and others are (worth to be) studied (such as micro-targeted ads, political activism, and the role of weak ties in network connections). Nevertheless, where you have been/are/will be says a lot about your activities, profile, and character – even in an offline life.

I have argued that the new features on Facebook are going beyond the idea of SNSs offering surveillance tools for imaginative traveling. These new features could trigger the audience/users not to be sufficiently satisfied with watching other friends traveling to exciting places, but to undertake this physical travel on their own. It can be functional or/and experience-focused trips – what is relevant and worthy of close study is the question if Facebook (as a platform that combines various services and application also of third parties) and corresponding online activities do lead to offline activities and increased travel mobility in particular.

I have mentioned other authors' argument that people need to physically travel to specific places in certain times in order to maintain their network connections (Urry 2003: 161; Urry 2007:36-37). If these connections are increasingly global through SNSs, mobility habits will change accordingly (cf. Larsen, Urry, and Axhausen 2006: 129; Sheller and Urry 2006: 221; Mau and Mewes 2009: 18). This poses multiple challenges on how we research these developments and how we re-/act on corresponding changes in practice. Our understanding of the interrelation between travel mobility and SNSs is still rather vague. With this paper, I have outlined which aspects of this interrelation can already be identified with the example of Facebook, and what other important questions can be addressed in future studies.

7. References

Adey, Peter, Lucy Budd, and Phil Hubbard (2007): Flying lessons: exploring the social and cultural geographies of global air travel. *Progress in Human Geography* 31 (6): 773-791.

Axhausen, Kay W. (2006): *Social networks, mobility biographies and travel: The survey challenges.* Arbeitsbericht Verkehrs- und Raumplanung 343. Institut für Verkehrsplanung und Transportsysteme/ETH Zürich: August 2006. Internet: http://e-collection.library.ethz.ch/eserv/eth:29514/eth-29514-01.pdf (21 March 2012).

Barkhuus, Louise, and Juliana Tashiro (2010): Student Socialization in the Age of Facebook. *CHI 2010 Proceedings of the 28th international conference on Human factors in computing systems.* New York (ACM): 133-142.

Burbary, Ken (2011): Facebook Demographics Revisited – 2011 Statistics. *Web Business Blog*: 7 March 2011. Internet: http://www.kenburbary.com/2011/03/facebook-demographics-revisited-2011-statistics-2/ (21 March 2012).

Cheung, Christy M.K., Pui-Yee Chiu, and Matthew K.O. Lee (2011): Online social networks: Why do students use facebook?. *Computers in Human Behavior* 27 (4): 1337-1343.

Ellison, Nicole B., Cliff Lampe, and Charles Steinfield (2009): Social Network Sites and Society: Current Trends and Future Possibilities. *Interactions* 16 (1): 6-9.

Facebook (2011a): A New Class of Social Apps on Facebook. *Official Facebook YouTube Channel*: 22 September 2011. Internet: http://www.youtube.com/watch?v=q3b94kFBah8 (21 March 2012).

Facebook (2011b): Interesting News, Any Time You Visit. *Official Facebook YouTube Channel*: 22 September 2011. Internet: http://www.youtube.com/watch?v=b6JrZdF4IPA (21 March 2012).

Facebook (2011c): Introducing Timeline -- a New Kind of Profile. *Official Facebook YouTube Channel*: 22 September 2011. Internet: http://www.youtube.com/watch?v=hzPEPfJHfKU (21 March 2012).

Facebook (2012): *Registration Statement on Forms S-1*. Internet: http://www.sec.gov/Archives/edgar/data/1326801/000119312512034517/d287954ds1.htm#toc (21 March 2012).

Fiveash, Kelly (2011): Facebook ditches Places – but embiggens location tracking. *The Register*: 24 August 2011. Internet: http://www.theregister.co.uk/2011/08/24/facebook_location_settings_places/ (21 March 2012).

Germann Molz, Jennie (2006): ‚Watch us wander': mobile surveillance and the surveillance of mobility. *Environment and Planning A* 38 (2): 377-393.

Ingram, Mathew (2011): Why Facebook's Frictionless Sharing Is the Future. *Bloomberg Businessweek*: 3 March 2011. Internet: http://www.businessweek.com/technology/why-facebooks-frictionless-sharing-is-the-future-10032011.html (21 March 2012).

Khisty, C. Jotin and Ulli Zeitler (2001): Is Hypermobility a Challenge for Transport Ethics and Systemicity?. *Systemic Practice and Action Research* 14 (5): 597-613.

Lanzendorf, Martin (2003): Mobility biographies. A new perspective for understanding travel behaviour. Paper presented at the *10th International Conference on Travel Behaviour Research in Lucerne*: 10-15 August 2003. Internet: http://www.ivt.ethz.ch/news/archive/20030810_IATBR/lanzendorf.pdf (21 March 2012).

Larsen, Jonas, John Urry, and Kay W. Axhausen (2006): *Moblities, networks, geographies*. Hampshire and Burlington (Ashgate).

Levy, Steven (2011): „Exclusive: Inside Facebook's Bid to Reinvent Music, News and Everything". *Wired Online – Epicenter*. 22 September 2011. Internet: http://www.wired.com/epicenter/2011/09/facebook-new-profile-apps/3/ (21 March 2012).

Mau, Steffen and Jan Mewes (2009): Class Divides within Transnationalism – The German Population and its Cross-Border Practices. In: Timo Ohnmacht, Hanja Maksim, and Max Bergman (Eds): *Mobility and Inequality*. Surrey and Burlington (Ashgate): 165-185.

O'Farrell, Renee (2012): The Problem of Mobility and Facebook's Battle for Revenue. *Insider Monkey Blog*: 8 February 2012. Internet: http://www.insidermonkey.com/blog/2012/02/08/the-problem-of-mobility-and-facebook%E2%80%99s-battle-for-revenue/ (21 March 2012).

Ohnmacht, Timo (2009): *Mobilitätsbiografie und Netzwerkbiografie: Kontaktmobilität in ego-zentrierten Netzwerken*. Dissertation. Philosophisch-Historische Fakultät/Universität Basel. Internet: http://edoc.unibas.ch/1008/1/final_phd-thesis-ohnmacht.pdf (21 March 2012).

Pring, Cara (2012): 100 social media statistics for 2012. *The Social Skinny Blog*: 11 January 2012. Internet: http://thesocialskinny.com/100-social-media-statistics-for-2012/ (21 March 2012).

Rhodes, Matt (2010): How social media is changing the way we travel. *Fresh Networks Blog*: 15 August 2010. Internet: http://www.freshnetworks.com/blog/2010/08/how-social-media-is-changing-the-way-we-travel/ (21 March 2012).

Shaffer, Justin (2011): Bigger, Faster Photos. *The Facebook Blog*: 16 November 2011. Internet: http://blog.facebook.com/blog.php?post=10150262684247131 (21 March 2012).

Sheller, Mimi and John Urry (2006): The new mobilities paradigm. *Environment and Planning A* 38 (2): 207-226.

Socialbrakers (2012a): *Facebook Page Statistics tagged as place*. Internet: http://www.socialbakers.com/facebook-pages/tag/place/ (21 March 2012).

Socialbrakers (2012b): *Facebook Page Statistics tagged as travel*. Internet: http://www.socialbakers.com/facebook-pages/tag/travel/ (21 March 2012).

Socialbrakers (2012c): *Facebook Statistics by country*. Internet: http://www.socialbakers.com/facebook-statistics/ (21 March 2012).

Socialbrakers (2012d): *Travel applications Facebook Statistics*. Internet: http://www.socialbakers.com/facebook-applications/category/65-travel (21 March 2012).

Szerszynski, Bronislaw and John Urry (2006): Visuality, mobility, and the cosmopolitan: inhabiting the world from afar. *The British Journal of Sociology* 57 (1): 113-131.

Urry, John (2002): Mobility and Proximity. *Sociology* 36 (2): 255-274.

Urry, John (2003): Social networks, travel and talk. *British Journal of Sociology* 54 (2): 155-175.

Urry, John (2007): *Mobilities*. Cambridge and Malden (Polity).

Wauters, Robin (2011): It's a Facebook World…Other Social Networks Just Live In It. *Tech Crunch*: 13 June 2011. Internet: http://techcrunch.com/2011/06/13/its-a-facebook-world-other-social-networks-just-live-in-it/ (21 March 2012).